FARM
MANAGEMENT
RECORD BOOK

Details

Name:	
Adress:	
E-mail adress:	
Phone Number:	
Fax Number:	

Log Book Details

Log Start Date:	
Log Book Number:	

LIVESTOCK RECORD

Year:		Start Date:			End of Year:			
#	Type of Livestock	Qty	AVG Weight	Value	Qty	AVG Weight	Value	Balance

Livestock Record X 34 pages

EQUIPMENT INVENTORY

#	Description	Purchase Date	Purchase Cost	Repairs	Current Value	Sale Price

Equipment Inventory X 4 pages

EQUIPMENT MAINTENANCE & REPAIR
MONTH:

Date	Equipment	Inspection/ Maintenance/Repair/ Services Required	Service/ Repair Date	Initials	Remarks

Equipment Maintenance & Repair X 20 pages

FARM EXPENSES
MONTH:

Date	Expenses	Cost	Remarks

Farm Expenses X 20 pages

FARM INCOME
MONTH:

Date	Source	Description	Method of payment	Amount

Farm Income X 20 pages

LIVESTOCK

RECORD

LIVESTOCK RECORD:

- Type of Livestock
- Quantity
- Average Weight
- Value
- Balance

LIVESTOCK RECORD

Year:		Start Date:			End of Year:			
#	Type of Livestock	Qty	AVG Weight	Value	Qty	AVG Weight	Value	Balance

Note: _____

LIVESTOCK RECORD

Year:		Start Date:			End of Year:			
#	Type of Livestock	Qty	AVG Weight	Value	Qty	AVG Weight	Value	Balance

Note: _____

LIVESTOCK RECORD

3

#	Type of Livestock	Qty	AVG Weight	Value	Qty	AVG Weight	Value	Balance

Year: **Start Date:**   **End of Year:**

Note: _____

LIVESTOCK RECORD

Year:		Start Date:			End of Year:			
#	Type of Livestock	Qty	AVG Weight	Value	Qty	AVG Weight	Value	Balance

Note: _____

LIVESTOCK RECORD

Year: _____ **Start Date:** _____ **End of Year:** _____

#	Type of Livestock	Qty	AVG Weight	Value	Qty	AVG Weight	Value	Balance

Note: _____

LIVESTOCK RECORD

Year:		Start Date:			End of Year:			
#	Type of Livestock	Qty	AVG Weight	Value	Qty	AVG Weight	Value	Balance

Note: _____

LIVESTOCK RECORD

#	Type of Livestock	Start Date:			End of Year:			
		Qty	AVG Weight	Value	Qty	AVG Weight	Value	Balance

Year: _____   **Start Date:** _____   **End of Year:** _____

Note: _____

LIVESTOCK RECORD

Year:	Start Date:			End of Year:			
#	Type of Livestock	Qty	AVG Weight	Value	Qty	AVG Weight	Value	Balance

Note: _____

LIVESTOCK RECORD

#	Type of Livestock	Qty	AVG Weight	Value	Qty	AVG Weight	Value	Balance

Year: _____ **Start Date:** _____ **End of Year:** _____

Note: _____

10 LIVESTOCK RECORD

Year:		Start Date:			End of Year:			
#	Type of Livestock	Qty	AVG Weight	Value	Qty	AVG Weight	Value	Balance

Note: _____

LIVESTOCK RECORD

Year:		Start Date:			End of Year:			
#	Type of Livestock	Qty	AVG Weight	Value	Qty	AVG Weight	Value	Balance

Note:_____

LIVESTOCK RECORD

12

Year:		Start Date:			End of Year:			
#	Type of Livestock	Qty	AVG Weight	Value	Qty	AVG Weight	Value	Balance

Note: _____

LIVESTOCK RECORD

Year: _____ **Start Date:** _____ **End of Year:** _____

#	Type of Livestock	Qty	AVG Weight	Value	Qty	AVG Weight	Value	Balance

Note: _____

LIVESTOCK RECORD

Year: _____   **Start Date:** _____   **End of Year:** _____

#	Type of Livestock	Qty	AVG Weight	Value	Qty	AVG Weight	Value	Balance

Note: _____

LIVESTOCK RECORD

Year: **Start Date:**   **End of Year:**

#	Type of Livestock	Qty	AVG Weight	Value	Qty	AVG Weight	Value	Balance

Note: _____

LIVESTOCK RECORD

Year:		Start Date:			End of Year:			
#	Type of Livestock	Qty	AVG Weight	Value	Qty	AVG Weight	Value	Balance

Note: _____

LIVESTOCK RECORD

Year: _____ **Start Date:** _____ **End of Year:** _____

#	Type of Livestock	Qty	AVG Weight	Value	Qty	AVG Weight	Value	Balance

Note: _____

LIVESTOCK RECORD

Year:		Start Date:			End of Year:			
#	Type of Livestock	Qty	AVG Weight	Value	Qty	AVG Weight	Value	Balance

Note: _____

LIVESTOCK RECORD

Year: **Start Date:**   **End of Year:**

#	Type of Livestock	Qty	AVG Weight	Value	Qty	AVG Weight	Value	Balance

Note: _____

LIVESTOCK RECORD

Year:		Start Date:			End of Year:			
#	Type of Livestock	Qty	AVG Weight	Value	Qty	AVG Weight	Value	Balance

Note: _____

LIVESTOCK RECORD

21

#	Type of Livestock	Qty	AVG Weight	Value	Qty	AVG Weight	Value	Balance

Year: **Start Date:** **End of Year:**

Note: _____

LIVESTOCK RECORD

Year:		Start Date:			End of Year:			
#	Type of Livestock	Qty	AVG Weight	Value	Qty	AVG Weight	Value	Balance

Note:_____

LIVESTOCK RECORD

Year:	Start Date:			End of Year:			
#	Type of Livestock	Qty	AVG Weight	Value	Qty	AVG Weight	Value	Balance

#	Type of Livestock	Qty	AVG Weight	Value	Qty	AVG Weight	Value	Balance

Note: _____

LIVESTOCK RECORD

24

Year:		Start Date:			End of Year:			
#	Type of Livestock	Qty	AVG Weight	Value	Qty	AVG Weight	Value	Balance

Note: _____

LIVESTOCK RECORD

Year:		Start Date:			End of Year:			
#	Type of Livestock	Qty	AVG Weight	Value	Qty	AVG Weight	Value	Balance

Note: _____

LIVESTOCK RECORD

Year: **Start Date:**   **End of Year:**

#	Type of Livestock	Qty	AVG Weight	Value	Qty	AVG Weight	Value	Balance

Note: _____

LIVESTOCK RECORD

#	Type of Livestock	Qty	AVG Weight	Value	Qty	AVG Weight	Value	Balance

Year: **Start Date:**   **End of Year:**

Note: _____

LIVESTOCK RECORD

28

Year:		Start Date:			End of Year:			
#	Type of Livestock	Qty	AVG Weight	Value	Qty	AVG Weight	Value	Balance

Note: _____

LIVESTOCK RECORD

Year: **Start Date:**   **End of Year:**

#	Type of Livestock	Qty	AVG Weight	Value	Qty	AVG Weight	Value	Balance

Note: _____

LIVESTOCK RECORD

| Year: | | Start Date: | | | End of Year: | | | |

#	Type of Livestock	Qty	AVG Weight	Value	Qty	AVG Weight	Value	Balance

Note: _____

LIVESTOCK RECORD

Year:		Start Date:			End of Year:			
#	Type of Livestock	Qty	AVG Weight	Value	Qty	AVG Weight	Value	Balance

Note: _____

LIVESTOCK RECORD

32

Year:		Start Date:			End of Year:			
#	Type of Livestock	Qty	AVG Weight	Value	Qty	AVG Weight	Value	Balance

Note: _____

LIVESTOCK RECORD

Year:		Start Date:			End of Year:			
#	Type of Livestock	Qty	AVG Weight	Value	Qty	AVG Weight	Value	Balance

Note: _____

LIVESTOCK RECORD

Year:		Start Date:			End of Year:			
#	Type of Livestock	Qty	AVG Weight	Value	Qty	AVG Weight	Value	Balance

Note: _____

EQUIPMENT

INVENTORY

EQUIPMENT INVENTORY:

- Description
- Purchase Date
- Purchase Cost
- Repairs
- Curent Value
- Sale Price

EQUIPMENT INVENTORY

#	Description	Purchase Date	Purchase Cost	Repairs	Curent Value	Sale Price

Note: _____

2

EQUIPMENT INVENTORY

#	Description	Purchase Date	Purchase Cost	Repairs	Curent Value	Sale Price

Note: _____

EQUIPMENT INVENTORY

#	Description	Purchase Date	Purchase Cost	Repairs	Curent Value	Sale Price

Note: _____

EQUIPMENT INVENTORY

#	Description	Purchase Date	Purchase Cost	Repairs	Curent Value	Sale Price

Note: _____

EQUIPMENT
MAINTENANCE & REPAIR

EQUIPMENT MAINTENANCE & REPAIR

- **Equipment**
- **Inspection**
- **Maintenance**
- **Repair**
- **Service Required**
- **Repair Date**

EQUIPMENT MAINTENANCE & REPAIR

		MONTH:			
Date	Equipment	Inspection/ Maintenance/Repair/ Services Required	Service/ Repair Date	Initials	Remarks

Note: _____

EQUIPMENT MAINTENANCE & REPAIR

2

		MONTH:			
Date	**Equipment**	**Inspection/ Maintenance/Repair/ Services Required**	**Service/ Repair Date**	**Initials**	**Remarks**

Note: _____

EQUIPMENT MAINTENANCE & REPAIR

MONTH:					
Date	Equipment	Inspection/ Maintenance/Repair/ Services Required	Service/ Repair Date	Initials	Remarks

Note:_____

EQUIPMENT MAINTENANCE & REPAIR

MONTH:

Date	Equipment	Inspection/ Maintenance/Repair/ Services Required	Service/ Repair Date	Initials	Remarks

Note: _____

EQUIPMENT MAINTENANCE & REPAIR

		MONTH:			
Date	Equipment	Inspection/ Maintenance/Repair/ Services Required	Service/ Repair Date	Initials	Remarks

Note: _____

EQUIPMENT MAINTENANCE & REPAIR

MONTH:					
Date	Equipment	Inspection/ Maintenance/Repair/ Services Required	Service/ Repair Date	Initials	Remarks

Note: _____

EQUIPMENT MAINTENANCE & REPAIR

MONTH:

Date	Equipment	Inspection/ Maintenance/Repair/ Services Required	Service/ Repair Date	Initials	Remarks

Note:_____

EQUIPMENT MAINTENANCE & REPAIR

MONTH:					
Date	Equipment	Inspection/ Maintenance/Repair/ Services Required	Service/ Repair Date	Initials	Remarks

Note: _____

EQUIPMENT MAINTENANCE & REPAIR

MONTH:

Date	Equipment	Inspection/ Maintenance/Repair/ Services Required	Service/ Repair Date	Initials	Remarks

Note: _____

EQUIPMENT MAINTENANCE & REPAIR

MONTH:					
Date	Equipment	Inspection/ Maintenance/Repair/ Services Required	Service/ Repair Date	Initials	Remarks

Note: _____

EQUIPMENT MAINTENANCE & REPAIR

MONTH:					
Date	Equipment	Inspection/ Maintenance/Repair/ Services Required	Service/ Repair Date	Initials	Remarks

Note: _____

EQUIPMENT MAINTENANCE & REPAIR

MONTH:					
Date	Equipment	Inspection/ Maintenance/Repair/ Services Required	Service/ Repair Date	Initials	Remarks

Note: _____

EQUIPMENT MAINTENANCE & REPAIR

MONTH:					
Date	**Equipment**	**Inspection/ Maintenance/Repair/ Services Required**	**Service/ Repair Date**	**Initials**	**Remarks**

Note: _____

EQUIPMENT MAINTENANCE & REPAIR

MONTH:					
Date	Equipment	Inspection/ Maintenance/Repair/ Services Required	Service/ Repair Date	Initials	Remarks

Note: _____

EQUIPMENT MAINTENANCE & REPAIR

MONTH:

Date	Equipment	Inspection/ Maintenance/Repair/ Services Required	Service/ Repair Date	Initials	Remarks

Note: _____

EQUIPMENT MAINTENANCE & REPAIR

MONTH:					
Date	Equipment	Inspection/ Maintenance/Repair/ Services Required	Service/ Repair Date	Initials	Remarks

Note: _____

EQUIPMENT MAINTENANCE & REPAIR

MONTH:					
Date	Equipment	Inspection/ Maintenance/Repair/ Services Required	Service/ Repair Date	Initials	Remarks

Note: _____

EQUIPMENT MAINTENANCE & REPAIR

MONTH:

Date	Equipment	Inspection/ Maintenance/Repair/ Services Required	Service/ Repair Date	Initials	Remarks

Note: _____

EQUIPMENT MAINTENANCE & REPAIR

		MONTH:			
Date	Equipment	Inspection/ Maintenance/Repair/ Services Required	Service/ Repair Date	Initials	Remarks

Note:_____

EQUIPMENT MAINTENANCE & REPAIR

MONTH:					
Date	Equipment	Inspection/ Maintenance/Repair/ Services Required	Service/ Repair Date	Initials	Remarks

Note: _____

FARM
EXPENSES

Annual Profit

YEAR:				
Month	**Description**	**Total Income**	**Total Expenses**	**Profit**

YEAR:				
Month	**Description**	**Total Income**	**Total Expenses**	**Profit**

FARM EXPENSES

MONTH:

Date	Expenses	Cost	Remarks

Note: _____

FARM EXPENSES

2

MONTH:			
Date	Expenses	Cost	Remarks

Note: _____

FARM EXPENSES

	MONTH:		
Date	**Expenses**	**Cost**	**Remarks**

Note: _____

FARM EXPENSES

	MONTH:		
Date	**Expenses**	**Cost**	**Remarks**

Note:

FARM EXPENSES

Date	Expenses	Cost	Remarks

MONTH:

Note: _____

FARM EXPENSES

	MONTH:		
Date	**Expenses**	**Cost**	**Remarks**

Note: _____

FARM EXPENSES

Date	Expenses	Cost	Remarks

MONTH:

Note: _____

FARM EXPENSES

MONTH:			
Date	Expenses	Cost	Remarks

Note: _____

FARM EXPENSES

	MONTH:		
Date	Expenses	Cost	Remarks

Note: _____

FARM EXPENSES

	MONTH:		
Date	**Expenses**	**Cost**	**Remarks**

Note: _____

FARM EXPENSES

Date	Expenses	Cost	Remarks
MONTH:			

Note: _____

FARM EXPENSES

	MONTH:		
Date	Expenses	Cost	Remarks

Note: _____

FARM EXPENSES

Date	Expenses	Cost	Remarks

MONTH:

Note: _____

FARM EXPENSES

	MONTH:		
Date	Expenses	Cost	Remarks

Note: _____

FARM EXPENSES

	MONTH:		
Date	**Expenses**	**Cost**	**Remarks**

Note:_____

FARM EXPENSES

MONTH:			
Date	**Expenses**	**Cost**	**Remarks**

Note: _____

FARM EXPENSES

	MONTH:		
Date	**Expenses**	**Cost**	**Remarks**

Note: _____

FARM EXPENSES

MONTH:			
Date	Expenses	Cost	Remarks

Note: _____

FARM EXPENSES

19

Date	Expenses	Cost	Remarks

MONTH:

Note:

FARM EXPENSES

	MONTH:		
Date	Expenses	Cost	Remarks

Note: _____

FARM
INCOME

Annual Profit

YEAR:				
Month	Description	Total Income	Total Expenses	Profit

YEAR:				
Month	Description	Total Income	Total Expenses	Profit

FARM INCOME

	MONTH:			
Date	Source	Description	Method of payment	Amount

Note: _____

2 FARM INCOME

				MONTH:
Date	**Source**	**Description**	**Method of payment**	**Amount**

Note: _____

FARM INCOME

3

Date	Source	Description	Method of payment	Amount

MONTH:

Note: _____

4 # FARM INCOME

			MONTH:	
Date	Source	Description	Method of payment	Amount

Note:

FARM INCOME

MONTH:				
Date	Source	Description	Method of payment	Amount

Note: _____

6

FARM INCOME

			MONTH:	
Date	Source	Description	Method of payment	Amount

Note: _____

FARM INCOME

MONTH:				
Date	Source	Description	Method of payment	Amount

Note: _____

FARM INCOME

8

Date	Source	Description	Method of payment	Amount

MONTH:

Note: _____

FARM INCOME

		MONTH:		
Date	**Source**	**Description**	**Method of payment**	**Amount**

Note: _____

FARM INCOME

			MONTH:	
Date	**Source**	**Description**	**Method of payment**	**Amount**

Note: _____

FARM INCOME

		MONTH:		
Date	Source	Description	Method of payment	Amount

Note: _____

FARM INCOME

12

Date	Source	Description	Method of payment	Amount

MONTH:

Note: _____

FARM INCOME

			MONTH:	
Date	Source	Description	Method of payment	Amount

Note: _____

FARM INCOME

MONTH:				
Date	Source	Description	Method of payment	Amount

Note: _____

FARM INCOME

		MONTH:		
Date	Source	Description	Method of payment	Amount

Note: _____

16 FARM INCOME

	MONTH:			
Date	**Source**	**Description**	**Method of payment**	**Amount**

Note: _____

FARM INCOME

MONTH:

Date	Source	Description	Method of payment	Amount

Note: _____

18 # FARM INCOME

			MONTH:	
Date	Source	Description	Method of payment	Amount

Note: _____

FARM INCOME

		MONTH:		
Date	Source	Description	Method of payment	Amount

Note: _____

FARM INCOME

			MONTH:		
Date	**Source**	**Description**	**Method of payment**	**Amount**

Note: _____

USEFUL CONTACTS/SUPPLIERS

NAME	PHONE NO.	DETAILS

USEFUL CONTACTS/SUPPLIERS

NAME	PHONE NO.	DETAILS

Notes

Notes

SUNDAY	MONDAY	TUESDAY	WENDSDAY	THURSDAY	FRIDAY	SATURDAY

MONTH:

SUNDAY	MONDAY	TUESDAY	WENDSDAY	THURSDAY	FRIDAY	SATURDAY

MONTH:

SUNDAY	MONDAY	TUESDAY	WENDSDAY	THURSDAY	FRIDAY	SATURDAY

MONTH:

SUNDAY	MONDAY	TUESDAY	WENDSDAY	THURSDAY	FRIDAY	SATURDAY

MONTH:

SUNDAY	MONDAY	TUESDAY	WENDSDAY	THURSDAY	FRIDAY	SATURDAY

MONTH:

SUNDAY	MONDAY	TUESDAY	WENDSDAY	THURSDAY	FRIDAY	SATURDAY

MONTH:

SUNDAY	MONDAY	TUESDAY	WENDSDAY	THURSDAY	FRIDAY	SATURDAY

MONTH:

SUNDAY	MONDAY	TUESDAY	WENDSDAY	THURSDAY	FRIDAY	SATURDAY

MONTH:

SUNDAY	MONDAY	TUESDAY	WENDSDAY	THURSDAY	FRIDAY	SATURDAY

MONTH:

SUNDAY	MONDAY	TUESDAY	WENDSDAY	THURSDAY	FRIDAY	SATURDAY

MONTH:

SUNDAY	MONDAY	TUESDAY	WENDSDAY	THURSDAY	FRIDAY	SATURDAY

MONTH:

SUNDAY	MONDAY	TUESDAY	WENDSDAY	THURSDAY	FRIDAY	SATURDAY

MONTH:

SUNDAY	MONDAY	TUESDAY	WENDSDAY	THURSDAY	FRIDAY	SATURDAY

MONTH:

SUNDAY	MONDAY	TUESDAY	WENDSDAY	THURSDAY	FRIDAY	SATURDAY

MONTH:

SUNDAY	MONDAY	TUESDAY	WENDSDAY	THURSDAY	FRIDAY	SATURDAY

MONTH:

SUNDAY	MONDAY	TUESDAY	WENDSDAY	THURSDAY	FRIDAY	SATURDAY

MONTH:

SUNDAY	MONDAY	TUESDAY	WENDSDAY	THURSDAY	FRIDAY	SATURDAY

MONTH:

SUNDAY	MONDAY	TUESDAY	WENDSDAY	THURSDAY	FRIDAY	SATURDAY

MONTH:

SUNDAY	MONDAY	TUESDAY	WENDSDAY	THURSDAY	FRIDAY	SATURDAY

MONTH:

SUNDAY	MONDAY	TUESDAY	WENDSDAY	THURSDAY	FRIDAY	SATURDAY

MONTH:

SUNDAY	MONDAY	TUESDAY	WENDSDAY	THURSDAY	FRIDAY	SATURDAY

MONTH:

SUNDAY	MONDAY	TUESDAY	WENDSDAY	THURSDAY	FRIDAY	SATURDAY

MONTH:

SUNDAY	MONDAY	TUESDAY	WENDSDAY	THURSDAY	FRIDAY	SATURDAY

MONTH:

SUNDAY	MONDAY	TUESDAY	WENDSDAY	THURSDAY	FRIDAY	SATURDAY

MONTH:

Made in United States
Orlando, FL
04 October 2024

52353226R00065